The Dinosaur Club
恐龙俱乐部

孵化恐龙

Dinosaurs Hatched

[英] 露丝·欧文/著

刘颖/译

汉英对照
恐龙科普

江苏凤凰美术出版社

全家阅读
小贴士

★ 每天空出大约10分钟来阅读。

★ 找个安静的地方坐下，集中注意力。关掉电视、音乐和手机。

★ 鼓励孩子们自己拿书和翻页。

★ 开始阅读前，先一起看看书里的图画，说说你们看到了什么。

★ 如果遇到不认识的单词，先问问孩子们首字母如何发音，再带着他们读完整句话。

★ 很多时候，通过首字母发音并听完整句话，孩子们就能猜出单词的意思。书里的图画也能起到提示的作用。

最重要的是，感受一起阅读的乐趣吧！

扫码听本书英文

Tips for Reading Together

• Set aside about 10 minutes each day for reading.

• Find a quiet place to sit with no distractions. Turn off the TV, music and screens.

• Encourage the child to hold the book and turn the pages.

• Before reading begins, look at the pictures together and talk about what you see.

• If the child gets stuck on a word, ask them what sound the first letter makes. Then, you read to the end of the sentence.

• Often by knowing the first sound and hearing the rest of the sentence, the child will be able to figure out the unknown word. Looking at the pictures can help, too.

Above all enjoy the time together and make reading fun!

Contents 目录

恐龙宝宝
Baby Dinosaurs

假如你生活在数千万年前，也许能
见到恐龙宝宝从蛋里孵化出来。
这些恐龙宝宝是慈母龙。

If you had lived tens of millions of years ago
you might have seen baby dinosaurs **hatching**
from eggs.
These baby dinosaurs are Maiasaura.

慈母龙蛋和你的脑袋一样大。
A Maiasaura egg was the size of your head.

慈母龙
Maiasaura
(my-uh-SOR-uh)

恐龙蛋
Dinosaur Eggs

所有雌性恐龙都下蛋。

科学家发现了岩石里的恐龙蛋。

但岩石里为什么会有恐龙蛋呢？

All female dinosaurs laid eggs.

Scientists find dinosaur eggs in rocks.

But how did the eggs get

into the rock?

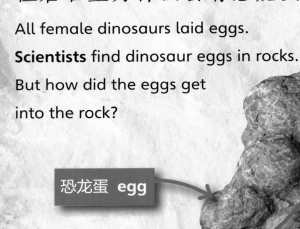

恐龙蛋 **egg**

岩石 **rock**

恐龙蛋有数千万年的历史。

A dinosaur egg is tens of millions of years old.

孵化中的霸王龙
a baby T. rex inside its egg

挖巢
Digging Nests

数千万年前，一些雌性萨尔塔龙来到河边挖巢。

Tens of millions of years ago, some female Saltasaurus dug nests by a river.

挖掘
digging

产卵
laying eggs

它们产下卵后就离开了。

它们没有等待自己的孩子出生。

They laid their eggs and went away.

They did not wait for their babies to hatch.

萨尔塔龙
Saltasaurus
(salt-ah-SORE-us)

每头雌性萨尔塔龙大约产卵25枚。
Each female Saltasaurus laid about 25 eggs in her nest.

寻找恐龙蛋
Finding Dinosaur Eggs

但是，泛滥的河水淹没了河岸，恐龙蛋被淤泥覆盖。

很久很久以后，它们变成了岩石。

数千万年后，科学家发现了岩石里的恐龙蛋。

But the river **flooded** and the eggs were covered in mud.

After a very long time they turned into rock.

Tens of millions of years later, scientists found the eggs in the rock.

恐龙蛋 eggs

萨尔塔龙蛋呈圆形，和甜瓜一样大。
A Saltasaurus egg is round and the size of a melon.

胚胎化石 A Fossil Baby

科学家从岩石里挖出恐龙蛋。

他们用X光观察恐龙蛋内部。

The scientists dug the eggs from the rock.

They used X-ray to look inside the eggs.

孵化中的萨尔塔龙模型
a model of a baby Saltasaurus in its egg

这些萨尔塔龙胚胎
是化石。
它们的骨骼已经
石化了。

The baby Saltasaurus were **fossils**.

Their bones had turned to rock.

萨尔塔龙有两辆小轿车那么长。

A Saltasaurus was as long as two cars.

围成圈的恐龙蛋
Circle of Eggs

有些恐龙会留下来照顾它们的蛋，直至孵化。

雌性葬火龙将产下的蛋围成一个圈。

然后它坐在中间，防止压碎这些蛋。

Some dinosaurs did stay with their eggs until they hatched.

A female Citipati laid her eggs in a circle.

Then she sat in the middle so she didn't crush the eggs.

围成一圈的恐龙
蛋化石
**a circle of
fossil eggs**

莽火龙看起来像一只巨大的鸟。

A Citipati looked like a giant bird.

莽火龙 Citipati
(SIT-i-patty)

15

化石巢穴
A Fossil Nest

科学家发现了一块化石，葬火龙正坐在一堆蛋上。

Scientists found a fossil Citipati sitting on some eggs.

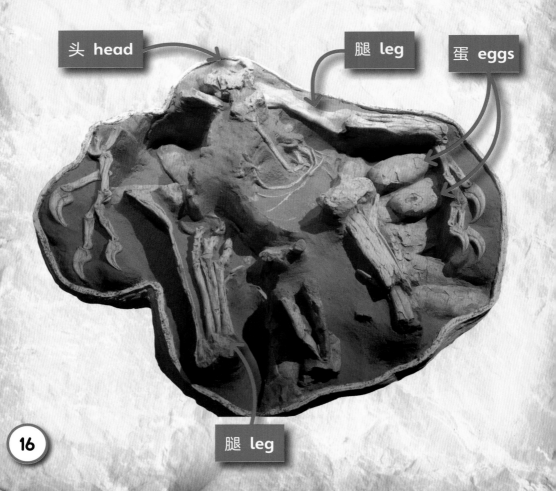

头 head

腿 leg

蛋 eggs

腿 leg

为什么葬火龙坐在蛋上？

也许是为了保持蛋的温暖和安全。

Why was the Citipati sitting on the eggs?

Maybe to keep them warm and safe.

雄性鸵鸟 male ostrich

蛋 eggs

这块化石可能来自一头雄性葬火龙。

有些雄鸟也会坐在蛋上，保持蛋的温暖和安全。

The fossil could be a male Citipati.

Some male birds sit on their eggs to keep them warm and safe.

许多巢穴
Lots of Nests

雌性慈母龙在泥土或沙子里筑巢。

每头雌龙在巢中产下30枚卵。

雌龙还将树叶和树枝盖在蛋上来保持温暖。

Female Maiasaura made nests in dirt or sand.

Each female laid 30 eggs in her nest.

She covered the eggs with leaves and twigs to keep them warm.

巢穴 nest

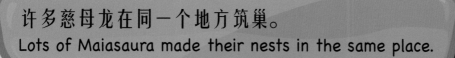

许多慈母龙在同一个地方筑巢。
Lots of Maiasaura made their nests in the same place.

雌性慈母龙
female Maiasaura

好妈妈
Good Mothers

雌性慈母龙会照顾它们的蛋和幼龙。

Female Maiasaura looked after their eggs and babies.

1岁大的慈母龙
one-year-old Maiasaura

慈母龙英文名的字面意思是"好妈妈蜥蜴"。

The name Maiasaura means "good mother lizard".

起初，幼龙不会走路，成年慈母龙便将植物带回去喂给它们吃。

At first their babies couldn't walk so they brought them plants to eat.

幼龙
babies

词汇表 Glossary

被……淹没 flooded

被大雨或上涨的河水漫过。

Covered by lots of water from heavy rains or overflowing rivers.

化石 fossil

存留在岩石中几百万年前的动物和植物的遗体。

The rocky remains of an animal or plant that lived millions of years ago.

孵化　hatching

从蛋里破壳而出。

Breaking out of an egg.

科学家　scientist

研究自然和世界
的人。

A person who studies
nature and the world.

恐龙小测验 Dinosaur Quiz

① 慈母龙蛋有多大？
How big was a Maiasaura egg?

② 萨尔塔龙蛋为什么被淤泥覆盖？
How did the Saltasaurus eggs get covered in mud?

③ 为什么雌性葬火龙将它的蛋围成一圈？
Why did a female Citipati lay her eggs in a circle?

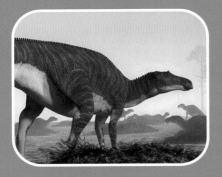

④ 慈母龙如何使它的蛋保持温暖？
How did a Maiasaura keep her eggs warm?

⑤ 为什么慈母龙被称为"好妈妈"？
In what ways were female Maiasaura good mothers?